WILDLIFE

HEDGEROW AND WAYSIDE

Alan Major

David & Charles
Newton Abbot   London   North Pomfret (Vt)

British Library Cataloguing in Publication Data

Major, Alan
 Hedgerow and wayside.—(Wildlife)
 1. Windbreaks, shelterbelts, etc.—Great Britain
 2. Natural history—Great Britain
 3. Roadside fauna—Great Britain
 4. Roadside flora—Great Britain
 I. Title   II. Series
 574.941   QH137

ISBN 0-7153-8185-7

© Alan Major 1982
Text and illustrations
All rights reserved. No part of this
publication may be reproduced, stored
in a retrieval system, or transmitted,
in any form or by any means, electronic,
mechanical, photocopying, recording or
otherwise, without the prior permission
of David & Charles (Publishers) Limited

Typeset by Typesetters (Birmingham) Limited
and printed in Great Britain
by A. Wheaton & Co., Exeter
for David & Charles (Publishers) Limited
Brunel House   Newton Abbot   Devon

Published in the United States of America
by David & Charles Inc
North Pomfret   Vermont 05053   USA

Contents

1 **Hedgerow and wayside habitat**
History, development, maintenance — 5

2 **Trees and shrubs**
Ash, elm, beech, hazel, holly, hawthorn, blackthorn, elder, privet, gorse, sycamore, wayfaring tree, spindle — 13

3 **Flowers, ferns and fungi**
Dog rose, ivy, honeysuckle, black and white bryony, bramble, woody nightshade, agrimony, herb robert, groundsel, cuckoo pint, creeping cinquefoil, jack-by-the-hedge, common polypody, orange-peel fungus — 20

4 **Birds**
Yellowhammer, greenfinch, goldfinch, treecreeper, nuthatch, tawny owl, chaffinch, robin, blackbird, dunnock, wren, blue tit, white throat, blackcap, pied wagtail, magpie, rook, kestrel — 33

5 **Mammals and reptiles**
Hedgehog, weasel, common shrew, bank vole, long-tailed field mouse, harvest mouse, viper, slow worm, common lizard — 40

6 **Insects and others**
Orange-tip, brimstone, small tortoiseshell butterflies, dragonflies, bumble bee, common wasp, crab spider, white-lipped banded snail, brown-lipped banded snail — 47

1
Hedgerow and wayside habitat

Hedgrows are one of the main features of the British countryside, and an attractive one we have come to take for granted. Their variety, the quality and quantity of their trees, shrubs and smaller plants, give lushness and interest to the landscape; for many wild creatures they are invaluable habitats, providing essential food, shelter, or both, not easily found elsewhere today.

Hedges of course are man-made, but their history goes back to the Stone Age, about 4000 BC, when the first people who can be called farmers cleared areas of woodland, established settlements and made ditch-surrounded enclosures for their goats, sheep and cattle. On top of the earth banks thrown up when the ditches were dug they added stockades of wooden stakes, sometimes lashing to them branches from nearby thorn trees, to make them more difficult for animals to penetrate. Thus the use of shrubs and trees for this purpose was established.

Through the Middle Ages and up to the seventeenth century, farming was usually a communal procedure, a large piece of land being cleared, often surrounded with a hedgerow, and divided into strips, each worked by its owner. As the open field method gradually changed, and especially when landowners began to enclose what had been common land, hundreds of miles of hedges were planted — the result being the patchwork landscape that we know today.

Hawthorn (*Crataegus monogyna*)

Double rows of hawthorn seedlings were used for most of the early hedgerows — hawthorn became an impenetrable barrier with regular trimming, and also it was easily propagated and grew quickly; hence its former name of 'quickset' or 'quickthorn'. Over the years other shrubs would grow up among it, and good shelter for livestock, crops and wildlife would build up.

A dictionary definition of a hedgerow is 'a fence formed of living trees or shrubs; a line of bushes, shrubs or small trees, usually hawthorn, privet, yew, etc, planted close, and trimmed level in height and cut back on both sides, so as to form a continuous fence or boundary, especially of a field or garden.' This is the usual idea of a hedge, whether the ornamental hedge planted for privacy in the front or back garden, the one bordering the side of a road, or those criss-crossing and dividing farmland. But there are many different forms.

The so-called 'wild' hedge bordering a lane or track is now usually a mixture of shrubs, sometimes with small trees and climbing plants. Through the survival of the fittest they combine to form the hedgerow. It may run rampant or be cut back on sides and top by the landowner, or by the authority responsible for the lane, when it begins to extend too far. The trees in it may be cut level or be cut around so they grow on in their natural shape. In this 'wild' hedgerow the trees may be there because they invaded it, perhaps from a nearby woodland, as seeds that germinated or as suckers from nearby trees. Elms, for example, throw up numerous suckers.

Other hedgerows began in this way and then were taken in hand by the landowner, shrubs and trees that will

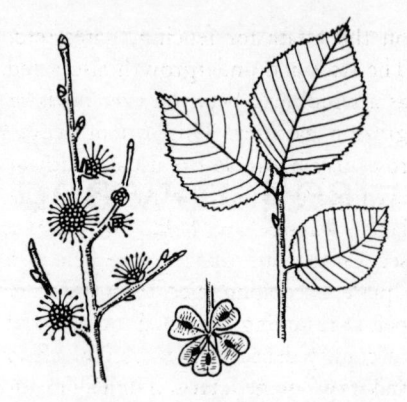

Common English elm (*Ulmus procera*)

not form a thick hedge being removed and suitable species being planted in their place. When there are species growing in the hedgerow that are not to be found wild in the nearby countryside this is the likely reason.

A hedgerow may now be all that survives of a former woodland that bordered the wayside. Much of our woodland has been felled and cleared, as the land has been wanted for crops or grazing. Clues to this having taken place are the plants in the hedgerow bottom. The presence of bluebells, wood anemones, yellow dead-nettle, wood sorrel and other woodland plants, sometimes primroses, indicates that woodland was cleared and the smaller trees and shrubs at its edge were kept as a hedgerow; new growth infilled it and people rapidly forgot the woodland was ever there.

Again, some of the present hedgerows are relics of small plantations of deliberately planted trees. Invading wild trees were sometimes allowed to grow and mature to provide timber, either to sell or to use on the estate for fencing, gates, etc. The trees and undergrowth also acted as a windbreak between two fields or grazing pastures. This sort of hedgerow may be treble the usual width or even more; it may be parallel to the lane, wayside or field, or have a section that juts out into the field as a square or oblong piece of hedgerow before resuming the usual form. Local or county dialect names were given to this type of hedgerow. In Kent, for example, 'carvet' is still used to describe a thick hedgerow or roadside copse; a 'shave' is a small fieldside copse or woodland, while a 'shaw' is a narrow plantation dividing two fields or a copse.

A narrow hedge, either 'wild' or deliberately planted, is often used to border a drainage ditch dug to take surface water from a thoroughfare or field. The hedgerow marks where the ditch is; it may be growing at the level of the adjoining land, or on top of the earth dug out for the ditch.

The hedgerow's form also depends on the landscape it is in, on local weather conditions, or the cultivation taking place and what it is required to do.

In Cornwall, Devon and Wales the 'hedgerow' may really be a bank of earth and rocks, specially laid as a facing, several feet high, bordering thoroughfares and intersecting fields and pastures. On the upper sides and top of the earth bank is a dense growth of blackthorn, hawthorn, furze, bramble or a mixture, perhaps with the occasional small tree in exposed areas; its stunted branches lean away from the dominant wind. All this is higher than the average person and affords shelter to the passerby or to livestock and crops in the enclosed fields or pastures. In Ireland, Wales and Scotland, rock may be used within banks of turf to strengthen it and slabs of rock used upright on the top of the bank. Plants and small shrubs establish themselves in front and between them. This differs greatly from a hedge in southern England, which may grow in richly fertile conditions as a surround to a ploughed field or pasture and consist of trimmed hawthorn or blackthorn, sometimes with holly or beech.

In a county such as Lincolnshire, where the cultivated land is in vast fields, the hedgerow may be a mere hawthorn strip lower than some of

Blackthorn (*Prunus spinosa*)

7

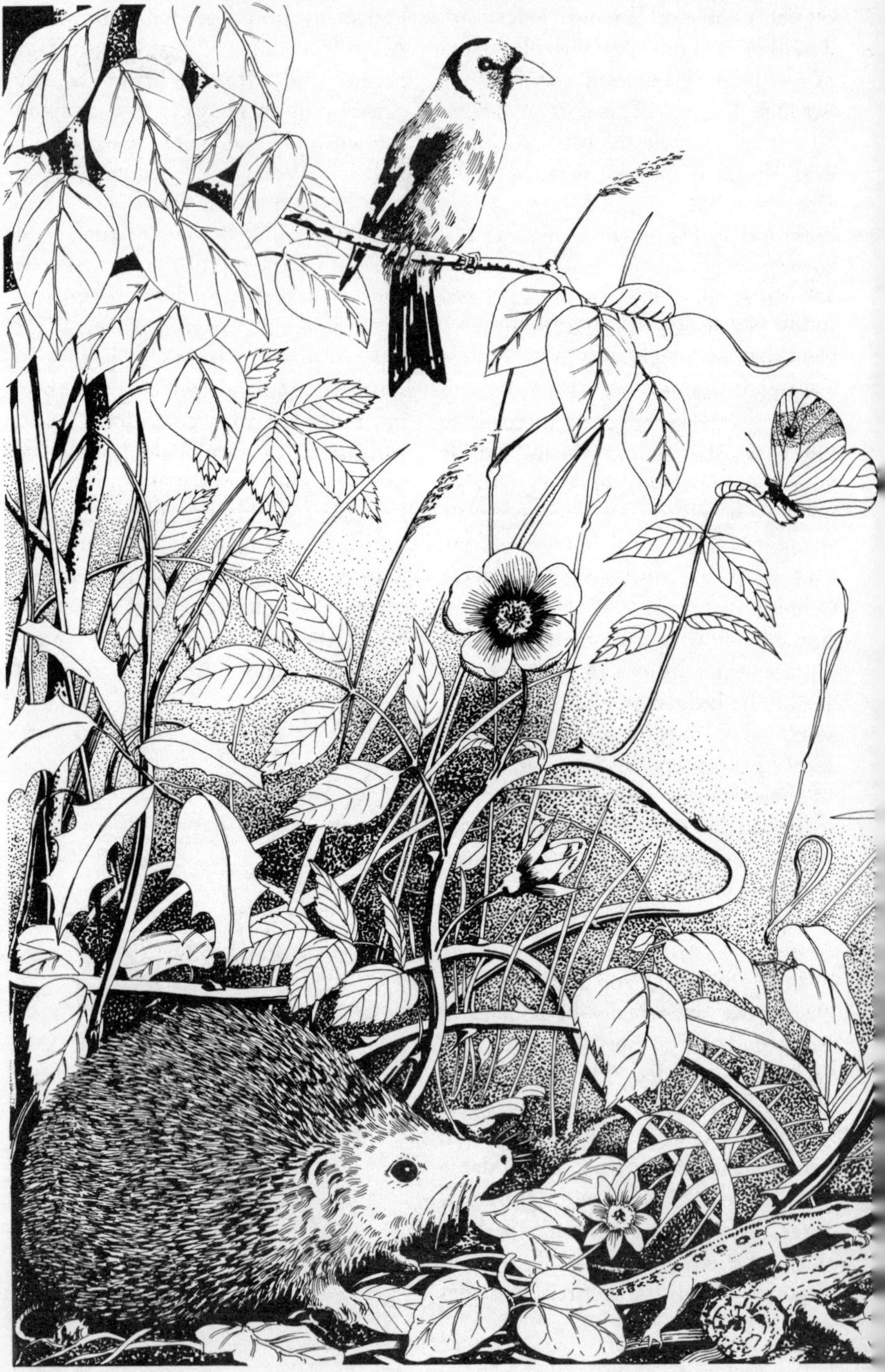

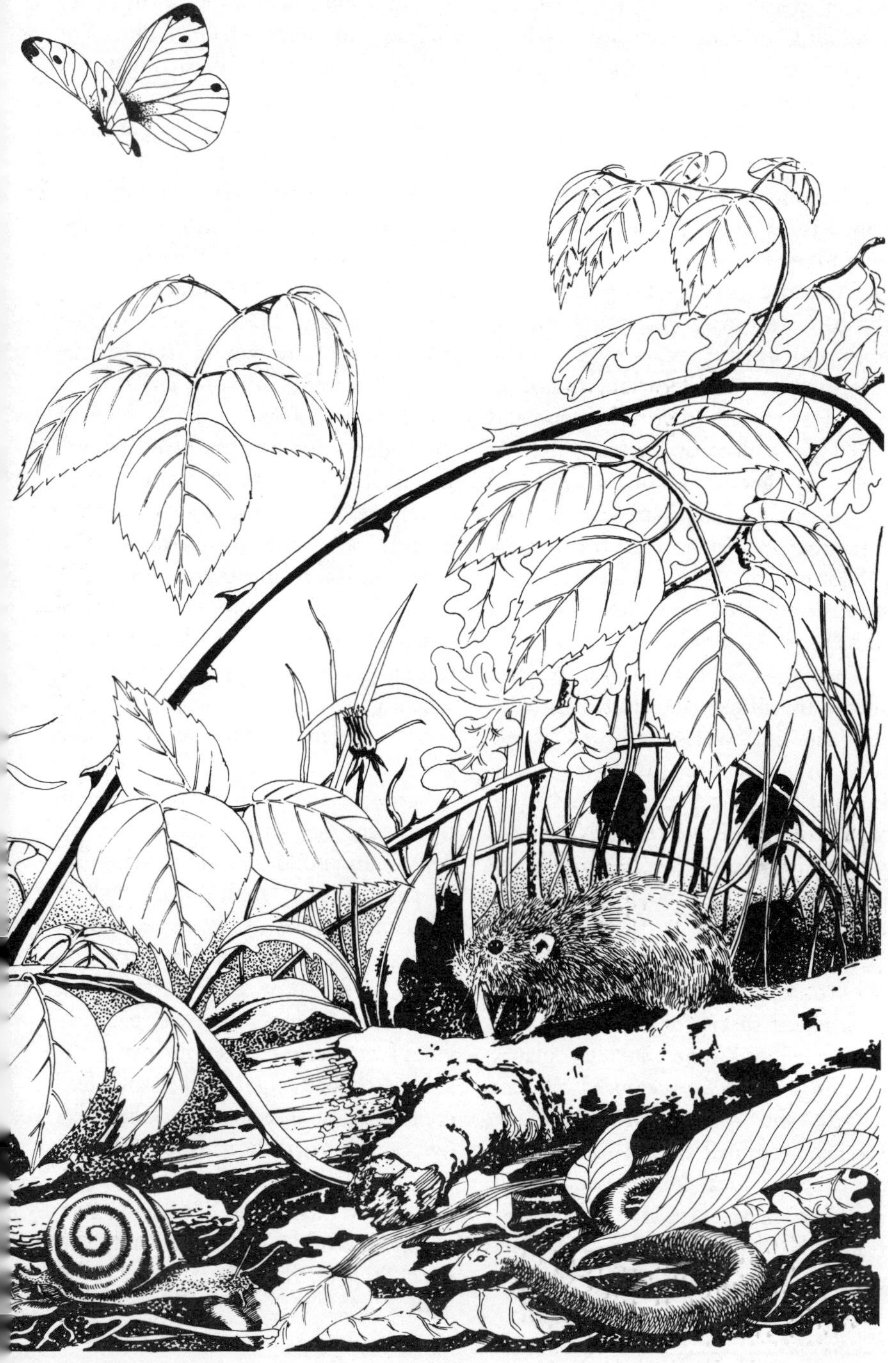

the plants alongside it. In Kent a hedgerow can be 12 feet tall or more, consisting of hawthorn trees which through generations of trimming and topping form a tight boundary and windbreak. Occasional enormous hedges can be seen: one is the famous beech hedge at Meikleour, Perthshire, which is 80 feet high and a third of a mile long.

The contents of the hedge, its trees, shrubs, flowering plants, animals, birds and insects, will vary of course according to their location. A hedgerow close to or bordering a woodland will have a flora and fauna different from the hedgerow on an exposed hilltop, in a grazing pasture, watermeadow or marsh near the sea. A hedgerow of fairly widely spaced shrubs will have less species than a tightly growing protective hedge.

The age of the hedgerow also affects the numbers of inhabitants. A centuries-old, well-established hedgerow will have been used by generations of creatures and plants. Many of them took to it when their original woodland homes were cleared, and ever since have found it a miniature nature sanctuary.

The age of a hedge can be calculated very roughly by counting the number of different species of shrubs and trees (not seedlings and smaller plants) growing along a measured 30-yard length, and allowing 100 years per species. Regional factors such as soil, climate and local hedge-management methods may of course affect your result, but generally speaking it has proved to be accurate within 100–200 years. If possible try to find documentary evidence to back up your findings: the local reference library or the county archivist's department may have maps, deeds or other documents recording dates of boundary changes or changes in land use.

As in the other habitats described in this series, in the hedgerow and country roadside an intricate web of life is balanced. The foliage of the various trees, bushes and smaller plants feeds the young stages, the larvae or caterpillars, of many insects; the dead leaf litter at the bottom of the hedge nourishes the plant growth and shelters other small creatures such as spiders, beetles and snails. All these in turn are food for certain birds; other birds eat the berries and seeds the hedge plants provide — berry-bearing shrubs such as hawthorn in fact crop more heavily in a hedge than when growing in a woodland edge where they are overshadowed by taller trees.

Some of the birds finding food in the hedge are also attracted by its density and build their nests there too, or roost there; larger birds, such as owls or magpies, a passing sparrowhawk, or predatory mammals such as stoats, will feed on small birds or their eggs and young. Kestrels pick up voles from verges; voles burrow in the soft earth of hedge bottom and feed on slugs and insects there; the slugs live on the vegatation. Rabbits, foxes and hedgehogs may make hidden homes for themselves. The hedgerow is a small world, each animal and plant depending on the others.

If the hedgrow is removed, the consequences for its inhabitants may be disastrous. The hedgerow-bottom soil with its leaf litter and debris will be ploughed in as part of the larger field; it is lost to the wild life. Many species there cannot move what would seem to them a great distance to the nearest similar hedgerow. Some animals and birds may be able to do so, but for the smaller creatures, the caterpillars of butterflies and moths, the ground beetles and so on, the plant species, especially those that need the permanently shady side of the hedgerow, it will mean certain extinction; and their disappearance diminishes the food supply for the others.

Hedgerow removal also has side-effects. Vipers, for example, hibernate in a hedgebottom in wintertime, especially where there is a dry bank and a tangle of tree roots; and the bank may be used as a lying-out place to absorb the sun's warmth in spring, after emerging from hibernation. Snakes may move to a nearby overgrown wayside or a scrubland waste for summer quarters, using the network of hedgerows as sheltered routes; other animals and birds too, if they need to leave one copse or woodland for another, will if possible follow the course of a hedgerow between the two sites, being much safer from predators on their journey. Acting as wind and snow breaks, hedges provide valuable shelter for crops and livestock. They slow evaporation from crops and reduce erosion of topsoil. In some of the English grain-producing areas in Lincolnshire and East Anglia, hedgerows stood in the way of using large farm machinery, but the result of their removal has been that the wind has blown away for ever the vital top soil in this open prairie-style landscape. Hedgerows also act as cover for game birds and habitats for wild bees needed to assist in pollinating clover, fruit trees and other crops.

The total mileage of hedgerow in Britain today is thought to be around 500,000, but going down fast. Several years ago the Nature Conservancy estimated that between 7,000 and 14,000 miles are removed *every year*. The main reason is the economics of modern farming. Smaller fields are sometimes awkward for large agricultural machinery, so the dividing hedgerows are removed to make one large field. Farmers sometimes consider hedgerows a stronghold for 'vermin', pests and weeds. Conservationists, on the other hand, say that the bird and animal life that depends on the hedgerow, apart from rats and mice, does little harm to cultivated crops and feeds on insect pests. Only three 'weeds', they claim, invade agricultural land from hedgerows — thistles, nettles and docks. And some species now live only in hedgerows; if the hedgerow goes, then the species disappears from the area.

But hedge maintenance is ever more costly, and it is now difficult to obtain skilled men to 'lay' the hedge in the traditional way. Without control the hedgerow becomes unmanageable, open and useless for its intended purpose, so farmers replace it with barbed wire or other fencing. Hedgerow and verge maintenance is costly for parish and rural councils, too. With busy traffic, even on narrow country roads, overhanging hedgerows can be a hazard, causing blind corners; so if the road is 'improved' the hedgerow is removed.

Hedgerows are also sometimes killed or badly damaged by careless spraying of nearby crops with chemicals which drift in the wind on to the foliage, or by herbicides used on road verges. They are burnt by stubble fires getting out of control.

But there are signs of public concern over the speed and extent of hedgerow destruction. Various research surveys have been carried out to discover more about what is happening. County nature trusts are using volunteers to restore or maintain old and interesting hedgerows. A few local authorities are also taking enough interest to avoid mechanical cutting of unusual rare shrubs and plants in a hedgerow, employing someone to do the job by hand. The Countryside Committees of some of the county councils and farmers' organisations sponsor hedgelaying competitions and courses to improve the standard of hedge laying and maintenance.

Mechanical hedge-cutters leave torn and splintered branches that would make the old-time hedgeman turn in his grave. Their speed, however, has reprieved many roadside hedgerows; without them, there is no doubt that thousands more miles of hedgerows would by now have been removed, for convenience and economy.

2
Trees and shrubs

John Evelyn, the famous seventeenth-century diarist, was also an arboriculturist who promoted the creation of hedgerows 'for the benefit and use of future generations'. For a quick hedge he recommended the hawthorn, because 'it provides a dense barrier of spiny twigs and its shape is easily kept in control'. After one year's growth he suggested that timber trees, such as ash, oak, beech or even fruit trees, should be planted at intervals in the hedgerow; two alternatives he advocated were dense blackthorn, which does not make quite such a 'tight' hedgerow, and elder, 'as every part of the tree is useful'. He also hoped to see 'the preciouser sorts of thorn and robust evergreen used more plentifully among the cornel and spindle and hips to make the hedgerow of a spiny hardihood'.

Hedge makers today would not necessarily agree with all he said, but estate owners and farmers in his lifetime benefited; they listened and they planted miles of hedgerows and plantation to make our countryside the patchwork panorama we enjoy. His words may well be behind some of the beautiful timber trees still seen in some old hedgerows — walnut, maple, oak, ash, fully developed in their natural shape, mature or nearly so, and never, or hardly ever, trimmed.

In the same hedgerow may be some of their offspring, saplings which have grown from seeds, fruit or suckers. Sometimes saplings have sprung up from the stump when a tree has been felled. This frequently happens with ash, elm and sweet chestnut.

The third part of the hedgerow mix are trees which now look like shrubs. When a tree sapling grows through the hedgerow shrubs and its main stem is cut back to the height of the remainder of the hedgerow, it then shoots out in all directions and with regular retrimming stays bushy.

Ash was formerly commonly grown in hedgerows when a tough elastic timber was required for making tool-handles, wheels and even hoops for barrels. It is common on limestone soils, and its pale grey bark, squat black buds and leaves with four to seven pairs of toothed leaflets make it quickly recognisable, summer or

13

winter. The buds are stout and the twigs thick due to the size of the leaves they support. Hedges often include some late shoots that have new dark purple leaves in summer. Mature ash trees produce thousands of winged seeds, known as 'keys'; most do not germinate, being consumed by seed-eating birds and animals. The ash is not a good tree for the hedgerow as its greedy, deep-penetrating rootstock starves other shrubs of minerals and water, so that they die back around it and make a gap.

Common or English elm has been the dominant hedgerow tree of the English Midlands and some of the South West; it was often planted at intervals in hawthorn hedges and, with its towering trunk and massive crown of branches it has become a traditional part of the English country scene. In Kent and East Anglia the smooth-leaved elm is more common. The winged seed is not usually fertile and the tree reproduces itself from numerous suckers around the trunk and from root-branches in the ground a considerable distance from the tree. These have to be cut out from a field, but in a hedgerow they help to maintain its density. An advantage of elm is that because of its height its billowing branches do not overshadow adjoining meadows and damage crop growth as other massive trees may do — although the branch wood is weak and twigs brittle, so they frequently create a litter on the ground. The double-toothed dark green leaves appear in early April in

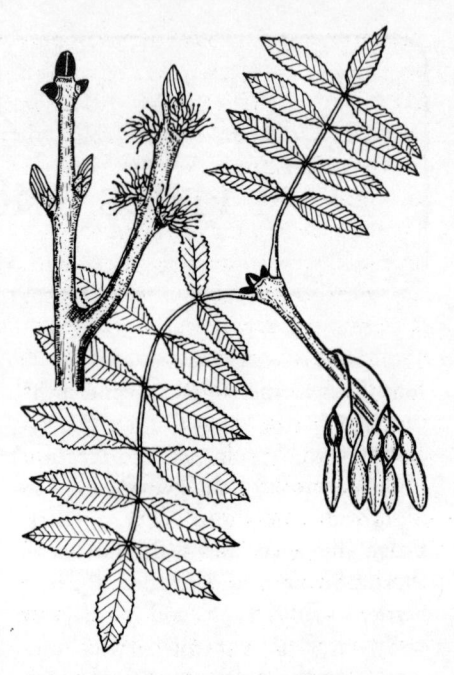

Ash (*Fraxinus excelsior*)

hedges, turning yellow, then gold, in late autumn.

In recent years Dutch elm disease, caused by a fungus carried by a beetle that feeds in the bark of dead or dying branches, has decimated the population of English elm trees, and left sad gaps in the hedgerows.

The **common beech** produces a magnificent umbrella of leaves that does cast shade over a field or wayside edge, restricting the growth of crops or nearby hedgerow. Yet this shade equally protects the soil from over-evaporation of moisture and when the leaves fall and rot they mulch and feed the soil. The beech will thrive on poor soil, and endures dry summers and severe winters. When it is planted and kept trimmed as a low hedge its twigs retain their dry brown leaves through the winter and act as a sheltering windbreak, as on the margins of Exmoor. The mature beech tree, with its massive smooth silvery trunk and huge dome of great branches, has a system of strong, spreading roots both above the soil and deep into it, so that sometimes you will see a huge beech with exposed roots that seems to be clinging precariously to the side of a hedgebank. In springtime the soft hairy leaves are pale green; in autumn a golden-brown that seems to set the beechwoods alight. The three-sided brown nuts, the 'mast', are enclosed inside a bristly case. Squirrels, mice and deer, tits and several other birds eat them eagerly, and the size of the mast crop can be vital to the survival of many of these creatures in winter. In the past it was fed to pigs.

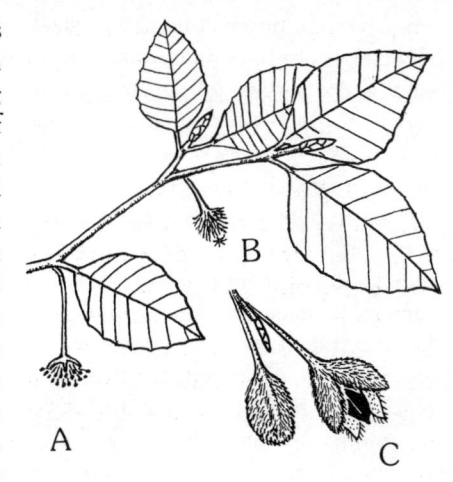

Beech (*Fagus sylvatica*)
(A) male flower
(B) female flower
(C) beechmast and case

Hazel is noticed on many a family walk, due to its hanging yellow 'lambs' tails', the male catkins, in late winter and its nuts in autumn. In the hedgerow it is commonest in the mixed hedges of the South West, Wales and the South East, along with other species such as maple and wayfaring tree, and not so common where most hedges are of planted hawthorn. It will remain a shrub if regularly cut back, but in open wayside becomes a tree up to 30 feet high. As with other catkin-bearing trees, the abundant pollen is distributed by the wind to the female bud-like flowers, which is why they need to appear early, before the dark green, coarsely hairy leaves

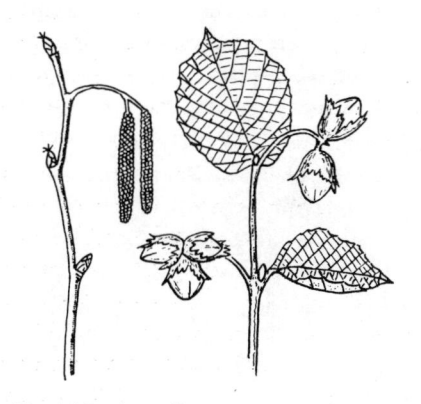

Hazel (*Corylus avellana*)

come out and obstruct its path. Hazel wood does not have much commercial value for furniture, but has long been used for gipsy pegs, walking sticks and fencing hurdles; a fork of hazel has been used for centuries by water diviners and long before bamboo canes were imported gardeners cut the suckers, dried them and used them to support their plants. The cob nut and Kentish filbert sold by greengrocers are varieties of hazel; its nuts are relished by birds such as nuthatches, by numerous animals — and by human beings.

Holly is unmistakable, famous as a Christmas decoration. When left to develop it becomes a tree up to 50 feet high, with smooth pale-grey bark. The leathery leaves on the lower portion are spiny, but towards the summit they may have a single spine at the tip or not even this. It was thought the plant had developed prickles to prevent deer and other animals browsing on the lower branches, but now it is thought to be a system of conserving water in the leaves by reducing their surface area. Holly is frequently used as an evergreen hedge, particularly in country house boundaries. When such a hedge has been trimmed for many years it becomes almost solid enough to walk along the top. Holly trees are either male or female, and the white, star-like female flowers must be pollinated from a male tree if they are to bear the scarlet berries. These may remain on the tree until the following spring if the winter is mild, but in hard weather birds will be grateful for them. The white timber, having grown slowly, has a fine grain and is very hard. It is used decoratively in furniture-making and for carving.

Hawthorn is one of our native trees and is common throughout Britain, not only deliberately planted to form hundreds of miles of hedgerows, but also growing as a thick bushy shrub or tree of the wayside. In the counties once extensively farmed on the old 'open field' system, especially in the East Midlands, it was the favourite hedging material when the big fields were divided up, and it remains the predominant hedge shrub. It will rapidly spread out from the hedgerow to invade neglected land and create scrub that changes the character of the area. Birds assist this by consuming the 'haws' and excreting the seeds some distance from the parent tree. Popularly known as the 'may tree', hawthorn was much used for May Day festivities, for decorating ex-

Holly (*Ilex aquifolium*)

teriors of houses, inns, churches and the maypole itself. Despite the widespread superstition that it is unlucky to take the beautiful creamy-white may blossom indoors, it was hung over doorways to prevent the entry of witches and evil spirits. The thorns are really a form of branch and sometimes are so long they bear leaves. There are two wild varieties in Britain — *Crataegus monogyna* and *Crataegus oxyacantha*, the Midland hawthorn. *C. monogyna* has smaller flowers and its leaves are more deeply indented, the branches are thornier and it is commoner in hedgerows and waysides. *C. oxyacantha* has larger white flowers and is more usually found at the edges of woodlands. Cultivated varieties have rose-pink or scarlet, single or double, flowers.

Blackthorn or **sloe**, related to the damson and plum, earns its name from the black bark on its main stem and twigs that makes a striking contrast to the welcome clusters of white, starlike flowers on the leafless twigs in March and April. If the weather was still severe when they appeared, country people said it was 'a blackthorn winter'. From the flowers the bitterly sour fleshy black fruit, or sloe, develops; it is sometimes made into wine — or sloe gin. The elliptical leaves were formerly dried to make a sort of tea. Whether in a hedgerow or growing separately, the blackthorn is rigid, with numerous tough branches turning in all directions armed with tough, sharp-tipped thorns. It often forms entire hedgerows or is mixed with hawthorn and other shrubs; also it will escape from a hedgerow and colonise derelict wayside.

Elder was sometimes planted as a hedgerow when a quick screen was wanted, but though a fast grower it would not make a stockproof hedge. Now it is usually seen as an ungainly, sprawling shrub in a hedgerow or as a many-branched small tree of the wayside. Its trunk and branches have a rough, corrugated corky bark, the green shoots quickly hardening into brown wood with a pith core that becomes hollow in old branches. The tiny white flowers are borne in flat-topped, dense clusters; they can be used for flavouring preserves or wine. The later dark purple-black berries are good for making wine, too. The leaves emit a noxious odour not to

Elder (*Sambucus nigra*)

everyone's liking, and bunches of them were hung in cottage windows to keep out flies.

Privet is easily recognised, so common is it as a garden hedge, able to withstand frequent trimming. When regularly cut back it bears green leaves through the winter; if a privet shrub is not given this treatment, many of the leaves fall in autumn and it bears exposed twigs with half-open shoots until springtime. The truly wild privet in a hedgerow or wayside has much narrower leaves than the cultvated sort. The single tubelike creamy-white flowers are formed in clusters, and the large purple-black berries are poisonous.

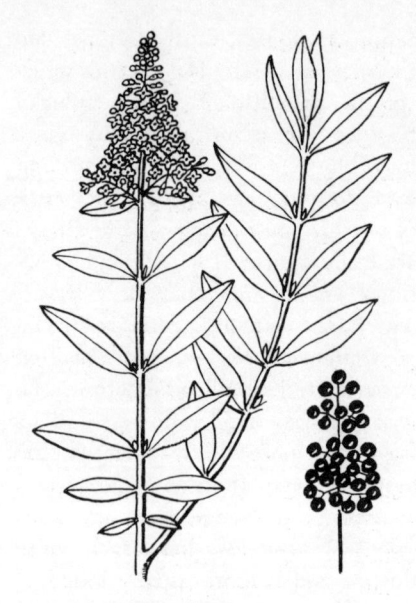

Privet (*Ligustrum vulgare*)

Gorse (or furze or whin) is an untidy evergreen shrub of some hedgerow banks or waysides, especially in heathy areas. Its inhospitable spiny foliage looks as if it would keep out virtually any man or beast; in fact, though, birds welcome it as a safe nesting site. The common gorse's large masses of showy, scented golden-yellow pealike flowers may be seen at any time of year, but especially in April to June; the smaller dwarf gorse flowers later. At first the tiny furze shrub has soft, hairy, trefoil leaves, but as it develops these are modified into spines. The larger spines are also the branches from which the flowers appear. They protect the plant against browsing animals and also prevent excessive loss of water through reducing the area of leaf surface; furze usually grows in

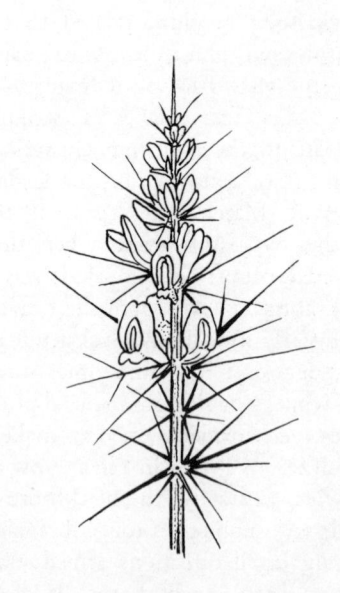

Gorse (*Ulex europaeus*)

exposed and heathy places with dry, shallow or stony soil. The flowers develop into hairy seed pods and on a hot day will burst with an easily heard crack, suddenly curling and flinging away the brown pealike seeds; furze soon ousts rival plants and colonises a site.

The **sycamore** or 'great maple' is not a native tree but was so widely introduced to our parks and woodlands in past centuries that it has become thoroughly naturalised. Both tall trees and regularly cut-down stumps, bursting with quantities of shoots, are common in our hedges. Where the sycamore has been allowed to grow, it can be a splendid tree with a dense, domed crown. The lobed leaves, however, are often spoiled by spots of black fungus. The yellowish flowers, hanging in bunches, are not particularly conspicuous, but on a biggish tree later in the year you cannot fail to see the bunches of paired, winged fruits. These are the seeds so often given to schoolchildren to draw as examples of fruits distributed by the wind!

The **field maple** is a genuinely British tree. It is smaller than the sycamore but closely related to it, and common in hedges except in the north. It may be found as a tree or trimmed down. The leaves are the same kind of lobed, palmate shape as the sycamore's, but smaller and with blunt, untoothed lobes; they are hairy underneath, as are the twigs, whereas the sycamore's are hairless. Maple fruits are also hairy, and their wings are straight and aligned, not angled as in the sycamore.

Wayfaring tree is to be seen by the wayfarer in hedgerows on chalky soils in the south of England. It will reach 20 metres high when unrestricted but is usually a mere 2 to 5 metres, growing best in dry places. The stems, leafless in winter, are never of tree proportions; the shoots are supple, slender, paired on the stem, and thickly coated with hairs; the wrinkled leaves being paired, too, and hairy underneath. As on the elder, the small white flowers are grouped together in flat heads, in May and June, succeeded by flattened bead-like red berries, which later turn black. In winter the large leaf buds have no protective bud scales and are covered with hairs to prevent frost damage.

The **spindle** tree is the straggly small tree or bush that has those brilliant pink fruits in autumn, which split open to show the bright orange-coated seeds. The smooth grey bark and small, elliptical leaves are less noticeable, but in areas with chalky or limestone soil this shrub is often an integral part of a hedge, untidy grower though it may be. Its hard wood was formerly used for spindles for spinning wool and for meat skewers.

3
Flowers, ferns and fungi

Like a well-planned garden a hedgerow or wayside has a continuance of flowering plants through the year, although obviously there are many more in the spring and summer months. Ferns, fungi and other forms of vegetation are also to be found.

About 600 different plants have been listed growing in hedgerows and waysides. The majoriy also occur in other places, such as woods or fields, so it is difficult to list 'typical hedgerow or wayside species', but there are many which are commonly found there. They may be using other plants, by climbing up them to the top of the hedgerow; they may grow at the hedgerow base, on the bank, or in a hedgerow or wayside ditch.

Hedgerows and waysides often offer plants several factors they need to thrive. Where they are left undisturbed during their growing and flowering period, perhaps being cut down once at the end of this, in late autumn or winter, they are able to produce seeds for next year's new plants or build up a store of food in their root system. If they have done the latter, some plants can burst into new growth almost before winter's end. Mechanical trimming of hedgerow verges may hinder their progress, but it may be done after the earlier-flowering plants have grown and seeded, and if done in high summer it allows plants time for further growth and seeding. Also the trimmer may only cut back the foliage of the verge and lower hedgebank near the road and not touch the plants higher up in the hedgerow.

Plants in a hedgerow are to a certain extent protected from low temperatures and the severe weather and winds that would knock them down in exposed places. The dampness of a hedgerow bottom is vital to some types of plants; tall, shady hedges have a humidity that plants such as black bryony require. Open hedgerows are required by wild roses and others with weak stems that need to lean on supporting stems.

Like trees and shrubs, flowering plants prefer particular types of soil. For this reason the flowers found in a hedgerow and wayside can vary widely in different counties; limestone has a rich flora generally, acid, peaty soils a more restricted one. Some plants tolerate almost any conditions.

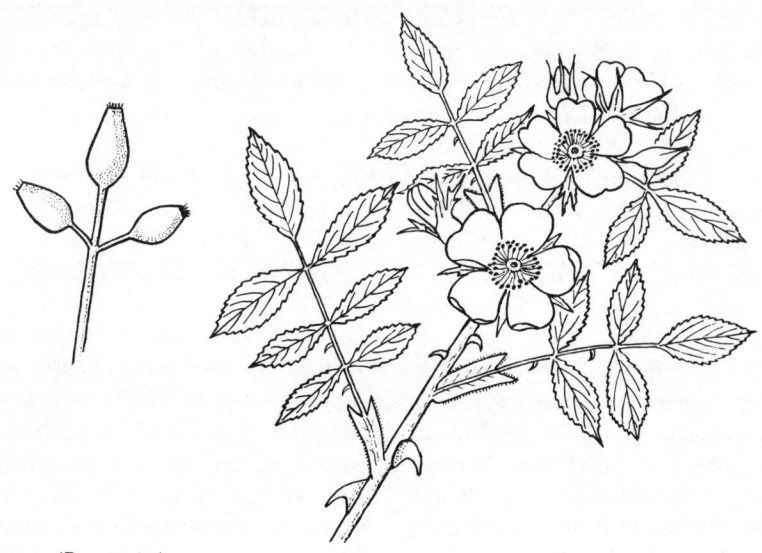

Dog rose (*Rosa canina*)

Dog rose, the largest British wild rose, and the commonest except in Scotland, uses the straight or hooked thorns on its stems to prop itself against neighbouring shrubs, its long branches climbing through the hedge. Its leaves are divided into 5–7 toothed leaflets. On open wayside or waste places it becomes a tall tangled bush. The pulpy section of the so-called 'hip' that ripens in autumn is really the swollen calyx part of the flower, the true fruits, hairy and nutlike, being inside it. Dog-rose hips are pitcher-shaped, which distinguishes them from other roses, on which the hips are rounded.

Ivy, the evergreen climber and trailer, uses the large numbers of adhesive rootlike claspers on its stems to secure itself to its host. As it winds its way up a hedgerow or wayside tree it often produces a huge mass of foliage and becomes a shrub, almost a tree, in itself, its thick stem nearly hiding the tree. It does not feed vulture-like on its host, as is often thought, but

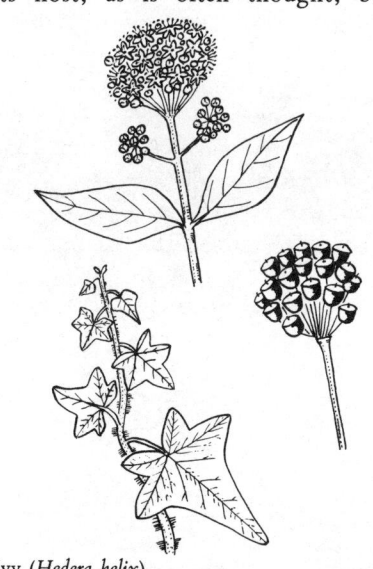

Ivy (*Hedera helix*)

21

manufactures its own food through its leaves and ground roots as other plants do. If the host tree dies it may be because the ivy has starved it of light and water. When the main stem of the ivy clinging to the tree is cut through the ivy will die. Ivy leaves vary in shape, most often being either the five-lobed shape found on non-flowering shoots, or the pointed heart-shaped type found on bushy flowering branches. The clusters of yellowish-green flowers appearing late, in October and November, are much sought by insects for the nectar and are often abuzz with flies, wasps, beetles and butterflies. The resulting black berries, though poisonous to man, are a valuable late-winter reserve food for birds such as wood pigeons, after the frosts have softened them. Some ivy never climbs but creeps along the hedgerow bottom and woodland floor and also never bears flowers.

Honeysuckle climbs up the side and to the top of the hedgerow by twining its long woody stem clockwise, from left to right, around the stems of neighbouring stronger plants. It may get up into hedgerow trees, or trail along and flop over the top of the hedge. The heads of long trumpet-shaped flowers, cream to buff, tinted with pink or red, protrude in all directions; in the evenings their strong, sweet odour lures passing long-tongued moths to pollinate them. The egg-shaped leaves, dropped in winter, are produed in opposite pairs on the stems. The clusters of bright shiny crimson berries, though eaten by birds, are poisonous to us. Bullfinches raid the berries to extract the seeds, the berry pulp being dropped to the ground where song thrushes and blackbirds consume it. A country name for the plant is woodbine, used also in places for the white convolvulus or bindweed.

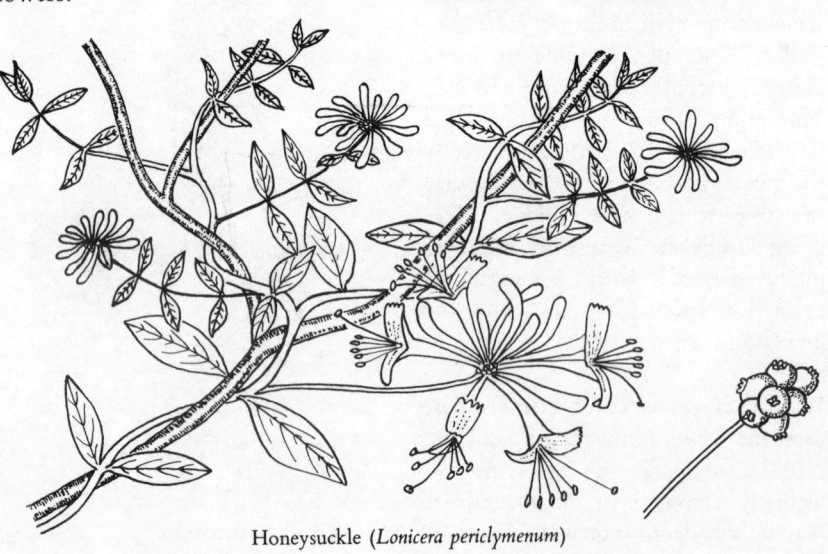

Honeysuckle (*Lonicera periclymenum*)

Black bryony (related to the yam) also climbs clockwise through a hedgerow. The long stems bear pointed, alternately growing, large, shiny, heart-shaped, dark green leaves on long stalks. These are conspicuous in many hedgerows in southern England. The small greenish flowers grow in loose spikes in May and June, and the clusters of soft scarlet berries are attractive in autumn and winter when the leaves and stems have become contrasting bronze or died back and dried out, but are poisonous to man. The plant survives the winter, the swollen fleshy root tubers 'hibernating' deep in the ground out of reach of penetrating frost. The roots are black on the outside, hence its name.

Black bryony (*Tamus communis*)

White bryony, in the cucumber family, is unrelated to black bryony, flowering May to September. The lovely green leaves are ivy-shaped, with five lobes; another difference is that the long stems haul themselves upwards in hedgerows by using tendrils that coil spirally around neighbouring twigs and plants. The flowers are greenish, dark-veined, and the small clusters of poisonous berries change from green to yellow, orange and finally bright scarlet in winter. The fleshy white root-tubers also 'hibernate' deep in the soil, then are able to send up shoots very rapidly in springtime.

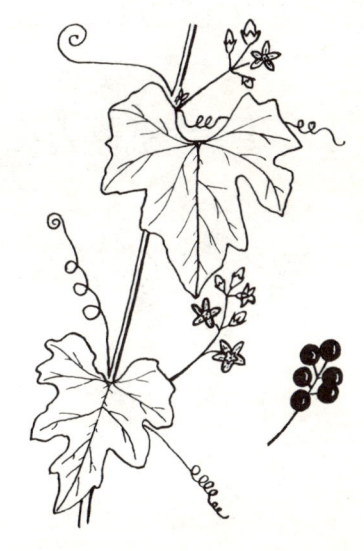

White bryony (*Bryonia dioica*)

Bramble (*Rubus fruticosus*)

Bramble or **blackberry** is another hedgerow scrambler and also often colonises large areas of wayside and waste land with its shrublike clumps; where its arching branches touch the earth they put down roots and a new plant grows, which in turn does the same thing. It also climbs, by using the hooky thorns on its thick stems and the backs of the leaves, familiar enough to all who gather ripe blackberries for jam or wine. Although popularly known as a berry the fruit is really made up of soft, fleshy, juicy drupes, each of which contains a seed. Birds eat the blackberries, fly elsewhere and excrete the seeds, which germinate where they fall on the earth and so the bramble spreads even further afield. The handsome pink or white flowers, attractive to butterflies, come out in June. There are several hundred varieties and hybrids.

Woody nightshade, or bittersweet, is far commoner than its infamous relative, deadly nightshade. It has weak, straggling downy stems several feet long which lean on stronger neighbours in the hedgerow and wayside, often clambering among brambles; dampish places suit it best. It is related to the potato and has similar-shaped flowers which have purple petals and yellow anthers forming a central tube. It flowers June to September. The very dark green leaves, stalked, are in two shapes: those lower down are often three-lobed, with one large and two smaller lobes; the upper stem leaves are spear-shaped. The fruits, egg-shaped berries which ripen from green to red, are poisonous, despite the country name 'bittersweet'.

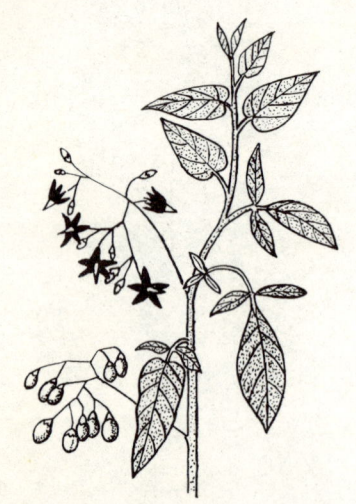

Woody nightshade (*Solanum dulcamara*)

(*opposite*) Woodland hedge, with the tall spike of a foxglove, formerly sought for the heart drug digitalis. Among the ferns the flowers of red campion can be seen (*John Beach/Wildlife Picture Agency*)

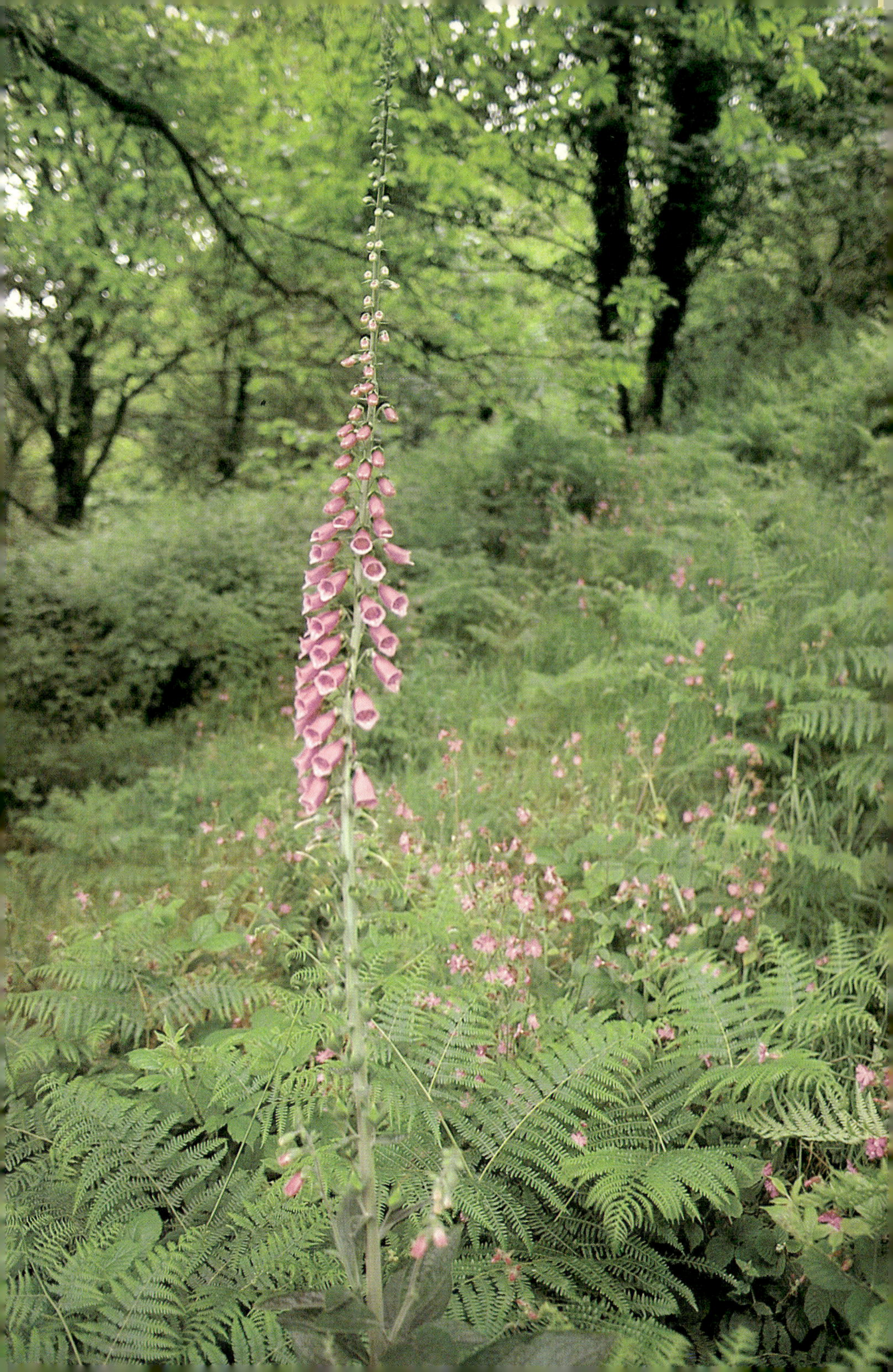

Roadside verge, with the yellow flowers of smooth hawk's beard, a relation of the dandelion. Mixed with it are the brown spikes and white flowers of lamb's tongue plantain (*Rodger Jackman/Wildlife Picture Agency*)

Roadside wall with fern in the crevices and lichen on the stones

Typical British field pattern, showing arable land and pasture divided by hedgerow, some containing maturing trees as well as shrubs (*Leslie Jackman/Wildlife Picture Agency*)

This hedgerow has a member of the parsley family (*foreground*) and the yellow flowers of charlock, which likes ploughed soil and survives in the hedgerow when cleared from fields (*John Beach/Wildlife Picture Agency*)

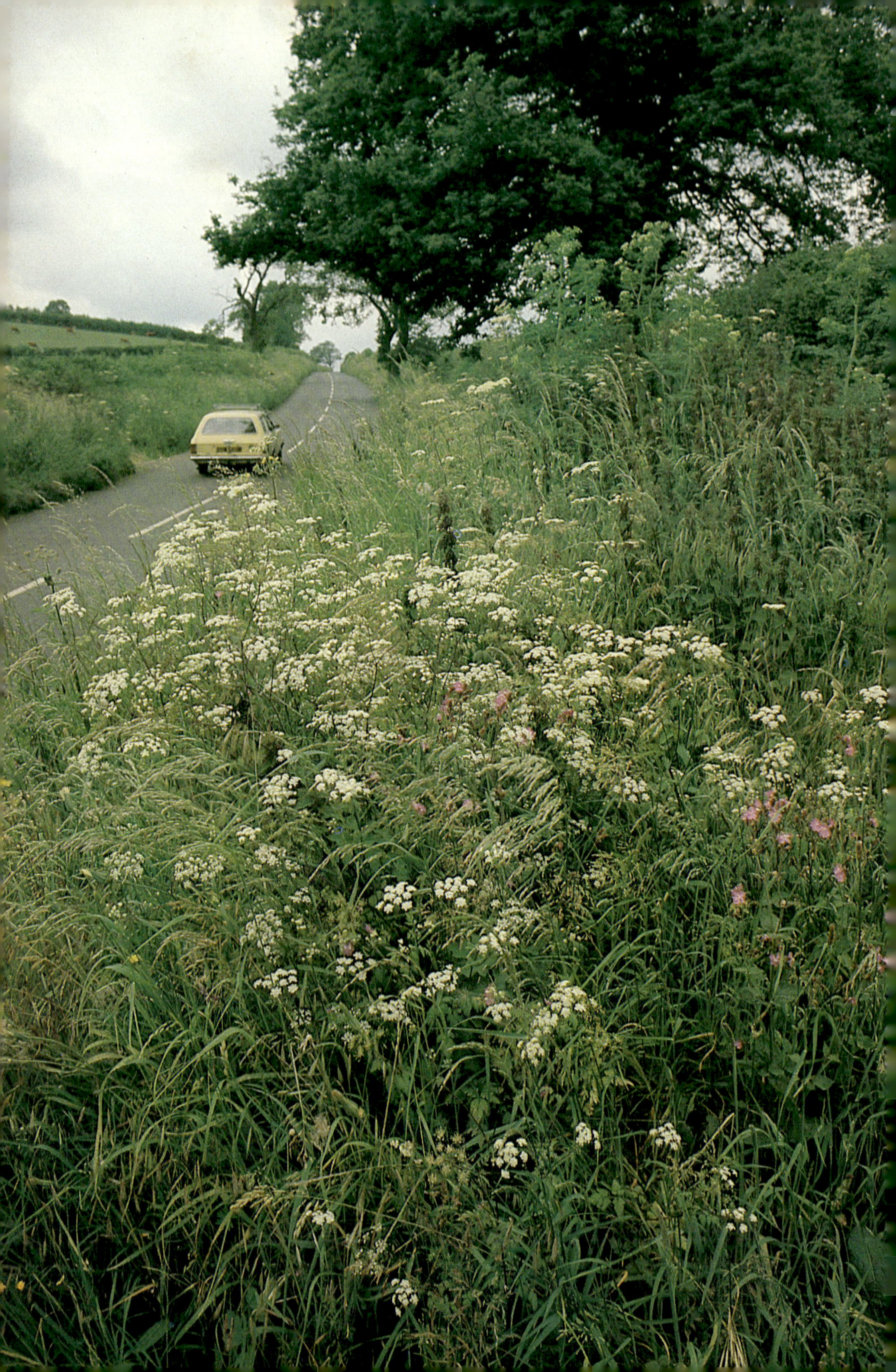

Agrimony is a pretty plant, found in grassy waysides and banks, flowering June to September. Its tall spikes of golden starry flowers on hairy stems give way to small hooked burrs, the fruits, which droop as they ripen and cling to passing animals or humans and thus get distributed elsewhere. The flowers, each of which stays open for three days, were once sought for the yellow dye they yielded, and the plant was used as a cure for liver ailments. The leaves are curious, being composed of seven to nine large, toothed leaflets which have between them several smaller ones on the same leaf stalk.

Agrimony (*Agrimonia eupatoria*)

Herb robert is another pretty but generally overlooked plant of the hedge-bank and waysides. Related to the geraniums or cranesbills, it is an annual, flowering May to October, with spreading, branching, hairy, bright-red stems and deeply divided ferny leaves. The pink flowers have darker red or purple streaks and are normally erect, drooping at night for protection from dew, or, in wet weather, from raindrops. The fruit, when not completely ripe, is pointed like a bird's beak. The plant, especially if trodden on, has a strong smell, which probably reduces its popularity with wild-flower gatherers.

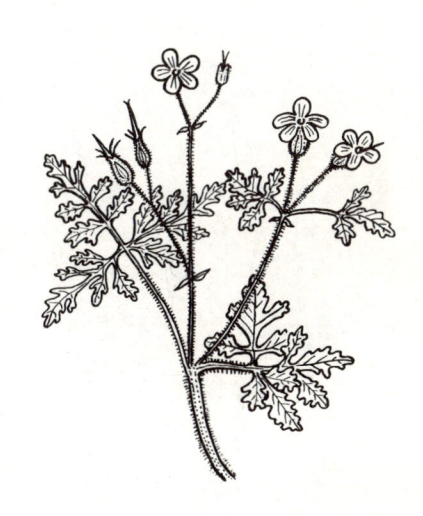

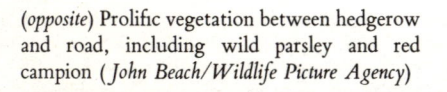

Herb robert (*Geranium robertianum*)

(*opposite*) Prolific vegetation between hedgerow and road, including wild parsley and red campion (*John Beach/Wildlife Picture Agency*)

Groundsel is one of the all-the-year-round flowers, familiar to everyone with a garden. On waysides and in hedgerow bottoms where adjacent land has been recently cultivated, it is a common 'weed', the wind taking the downy seeds to new sites or spreading them around the parent plant. It has some uses though: rabbits enjoy it and birds both wild and caged like the seeds. The leaves are alternate, clasping the stem.

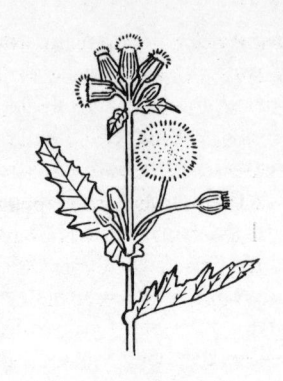

Groundsel (*Senecio vulgaris*)

Cuckoo pint (known as lords and ladies, or wild arum) in the hedgerow bottom may not look like a flowering plant. The male and female flowers are hidden in the bulbous part above the green stem, April to June. Some more undeveloped male flowers are immediately above them around the base of a conspicuous long, club-like purple structure called the spadix; flowers and spadix are backed by a broad pale-green hood, rather like a monk's cowl, called the spathe. The flower emits a foul, urine-like (some say) smell that attracts flies. These crawl down the spathe, past the undeveloped male flowers, past the true male and female flowers and

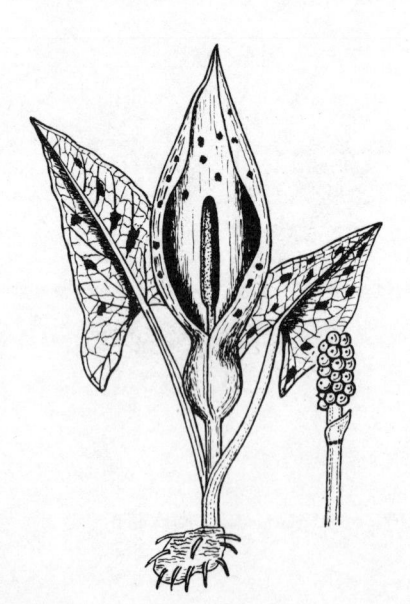

Cuckoo pint (*Arum maculatum*)

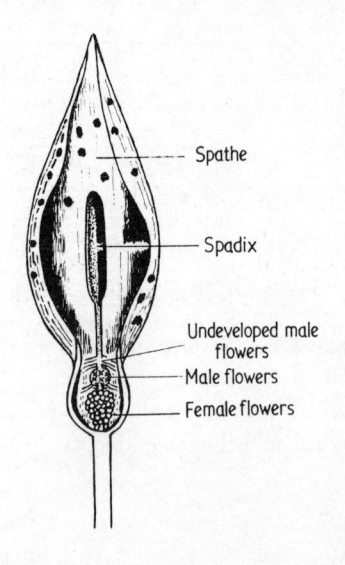

30

wander about over the female flowers, pollinating them. The snag for the flies is that when they try to find their way out again the downward-pointing undeveloped hairy male flowers prevent them; they are trapped there until all the female flowers have been pollinated. Then the upper hairy male flowers wither and the flies escape. But they usually go to another cuckoo-pint flower, to go through the same pollination experience again. From the female flowers develop the vivid orange-red poisonous berries, handsomely clustered in a spike. The long-stalked, large leaves are arrow-head-shaped with purple and black spots. The white tubers and roots used to be dug up for their starch, but are now considered poisonous. Cuckoo pint has numerous country names — cow-and-calves, parson-in-the-pulpit, priest's pintle, calf's foot.

Creeping cinquefoil is to be found along the hedgebottom and roadside verge, where it is able to survive the dust and dirt. The yellow flowers, seen from June onwards, look something like long-stalked buttercups although not as cup-shaped; the cinquefoil is closely related to the wild strawberry and belongs to the rose family. From the parent plant, slender stems spread in various directions, sometimes being yards long, at intervals bearing the palmate leaves — the five-toothed leaflets are arranged finger-fashion. Roots grow from the underside of the stems at intervals, to anchor them down, and from these points a new plant grows.

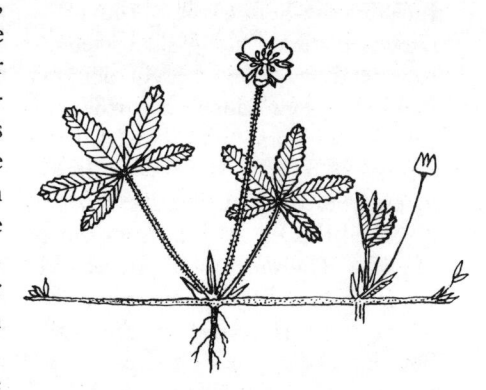

Cinquefoil (*Potentilla reptans*)

In shady, damp banks and hedgerow waysides some of our most beautiful ferns will flourish. Some of the smaller species like a niche among rocks jutting from the bank or the mossy, decaying stump of a tree.

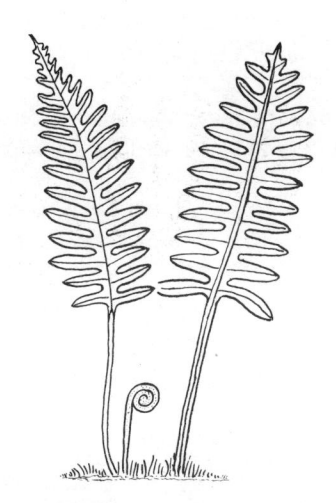

Polypody (*Polypodium vulgare*)

Jack-by-the-hedge, or garlic mustard, is common along shady hedgerows and waysides. Rarely do you find one plant alone; it grows in colonies of sometimes hundreds of plants. It is erect, tall, with fresh-looking heart-shaped, toothed foliage; if rubbed this gives off a strong smell of garlic. The small white flowers are out in spring, followed by cylindrical seedpods. It is related to the wallflowers.

Common polypody occurs among the leaf mould in old hedgerows, or its fleshy roots find a hold in rotting leaves up in the forks of tree branches. The evergreen, leathery, oval-oblong fronds, 18 inches long, are deeply cut into lobes. The better the growing conditions the larger the fronds will be. They are evergreen and when they become well-established their luxuriant growth may adorn the hedgerow bank for years. The brownish-orange spore cases are in rows along the leaflets.

Fungi to be found along hedgerows and waysides are feeding on rotting matter in the soil or upon the decaying tissues of unhealthy or dead trees or shrubs.

Orange-peel fungus is unmistakable in the hedgerow bottom or on the exposed earth of a hedgebank: it looks just like a piece of orange peel thrown on the ground, bright orange on top and paler underneath. Edible, up to 5 inches wide, it is attached to the ground by a tiny stem. In a favourable place several may be growing close together so that it is difficult to distinguish each separately. It is also fragile and cannot be pulled up, breaking into pieces in the fingers. It occurs in late summer, autumn or early winter.

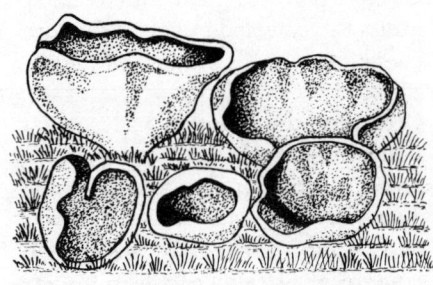

Orange-peel fungus

4
Birds

Large numbers of different birds fly into and out of hedgerows and waysides. Many are only 'passing through', others live in the hedgerow or nearby, and a few species you are sure to see before long if you watch.

Birds use hedgerows and waysides for various purposes. The hedge is a place to perch while observing the surrounding countryside for food, or enemies, or for others of its own kind if it is a species that lives its life in flocks, or while calling or singing. For some birds it is a good place to obtain food, either on the shrubs or plants or among the debris at the hedgerow bottom or by the wayside, and for these birds the hedge is a convenient place in which to build a nest and raise young. Hedgerows also give shelter to birds, concealment from predators in the air such as hovering birds of prey, and a place to roost at night; owls roost in tall hedgerow trees by day. It is known that many woodland-living birds will adapt to live in the narrower confines of a hedgerow or a wayside with a number of trees if their wood is felled — and most of the birds we think of as hedgerow species in fact adapted to that habitat when woodland was cleared in the past. This is one reason why hedges are now such a valuable habitat.

If the hedgerow trees are high enough rooks will create a small colony of nests, while a single pair of carrion crows or wood pigeons may nest in a hedgerow with just a few trees. If a tree is hollow and has a hole that can be adapted or enlarged, little owls, barn owls, stock doves or starlings will nest there.

The species of birds and their numbers vary according to the food supply on the trees and shrubs — insects and their larvae, seeds, fruit and berries — and the slugs and snails, worms, and mice and other rodents at the base. For this reason hedgerows that are regularly trimmed and have the bottom cleared out occasionally, or cut back, shelter fewer birds and other creatures than those that are neglected and luxuriantly overgrown. Certain hedgerow and wayside trees, shrubs and plants are hosts to more insects than others. The wayside oak is known to be a source of food to over 500 different insects. A considerable number live on the hazel, but less on the beech and practically none on sweet and horse chestnut.

The **yellowhammer** or **yellow bunting** is a typical top-of-the-hedgerow bird, the male being especially conspicuous with its bright yellow head and underparts, the female is less yellow. It perches on a prominent hedge branch surveying the open country and uttering its unmistakable song, 'a little bit of bread and noooooh cheese'. Yet the nest of grass, moss and stalks, lined with hair obtained in the vicinity, is built near the ground in the bottom of the hedgerow or a shrub. It feeds chiefly on seeds from the hedgerow and adjoining countryside.

The **greenfinch** also sings from a jutting branch but usually higher up, in a tall shrub or upper part of a hedgerow tree, where it also builds its nest. It is the largest of the finches, has a rather solid 'muscular' appearance, and olive-green colouring with bright yellow feathers on the male's wings and tail. It also feeds mainly on seeds but is attracted to a bird table offering peanuts and sunflower seeds.

The **goldfinch** is immediately recognisable, with its bright red face, black cap, white sides to the head and golden-yellow wing-bars. It heralds its arrival in the hedgerow top or on the seeding thistles of the wayside with its cheerful tinkling notes, similar to a canary's. Usually there is a party of goldfinches feeding together, except in the breeding season that

Goldfinch (*Carduelis carduelis*)

starts in May, their dancing flight being another attractive characteristic. The bill is rather more pointed than that of most finches, the goldfinch being thus equipped to extract the seeds deep in teasel or thistle heads. In recent years goldfinches have greatly increased in numbers; they are now protected by law from the cruel and once-common fate of being captured and sold as singing cage-birds.

The **treecreeper** may be one of those 'little brown jobs', but it is easier to identify than most of the small brownish birds that flit around the hedgerows, because of its behaviour — and its long toes, if you are close enough to see them. It looks as much like a mouse as a bird, as it climbs up the trunk of a tree or runs along the underside of a branch, searching for insects and grubs in the bark. Treecreepers are not often seen in groups, but they do sometimes join flocks of tits and goldcrests in winter. With their whitish underparts and high-pitched burst of song, they may be quite often noticed in hedges with trees, and are particularly interesting to watch.

The **nuthatch** is another smallish bird that runs up treetrunks — or down, or sideways, upside-down or any way — but with jerky movements and much more conspicuously than the treecreeper. Its appearance is quite different too: its chunky, squat shape, blue-grey back and pinky-buff underparts are handsome and distinctive. Its loud tapping as it opens the acorns and nuts it wedges in crevices often lead you along to look for it. The nuthatch nests in a cavity in a tree, perhaps a knot-hole or an old woodpecker hole, and will reduce the size of the entrance by filling it in with wet mud.

The **tawny owl** (or brown owl) is the largest common British owl, living mainly in well-wooded areas though occasionally right in town. You are most likely to notice it first by its eerie call, 'hooo, hooo, hoo -oo-oo', or the sharp 'kee-wick' that often answers the hoot; you may be lucky enough to see it hunting in broad daylight, looking for small rodents at the base of the hedge or in the verge. It most often roosts in a hollow tree or bush by day, but a roosting owl can sometimes be found by following up a noisy group of smaller birds that seek it out and mob it.

The **chaffinch** is related to the former two finches and in areas with

Chaffinch (*Fringilla coelebs*)

plenty of trees and bushes, including either rural or town gardens with hedgerows, it is often the commonest bird. In winter it joins large flocks of other finches. The male has a slate-blue crown and nape, black forehead, red back, green rump and pinkish underparts. During its undulating flight from place to place, the white bars on the wings and white patches on the outer tail feathers are conspicuous, the male's colouring being generally brighter than the female's. When living close to man it can become quite tame, and its distinctive repeated 'pink pink' call note identifies it immediately. It builds a beautiful cup-shaped nest low down in hedgerows, bushes and small trees, using grass, moss and fibres, lines it with feathers, wool and hair and adorns it with lichen, flakes of tree bark and spiders' webs that blend the nest with its surroundings.

The **robin**, a familiar sight in the hedges and shrubberies of parks and gardens, is also a common resident of the rural hedgerow and wayside. The male and female have the same colouring but the young have mottled brown and buff plumage. Universally loved, the Christmas-card symbol of winter festivity, tame enough to fly down picking up insects and grubs around your feet as you work on the garden soil, the robin is far from friendly to its own kind. Each robin has its own territory, an area sufficient in food to support itself and its mate, and woe betide any other robin that ventures into its patch; it will even maim or kill a too-persistent invader. In a breeding season when males are scarce, however, each one may have a 'harem' of several mates in a larger territory, or two mates rather closer together, dividing his time between them. He sings throughout most of the year . . . as a warning to others to stay out. The piping notes of autumn, issuing from deep in a hedgerow on a misty day, have a peculiarly melancholy quality.

The **blackbird** is often the noisiest bird in the hedgerow as you pass near. The presence of a snake or other predator, or the approach of man, often causes the male bird to sound a warning to all and sundry in the area. He retreats into a hedgerow bottom or under a bush, making a chattering alarm, a 'chak-chak' or 'chink-chink'. This is often uttered at roosting time as dusk falls, a dangerous time of day for all perching birds; if a danger gets close he becomes almost hysterical, making a metallic screeching from deep inside the hedgerow. But when singing, often from the vantage point of a tree top or hedge branch, the blackbird is probably Britain's foremost song bird, with loud, rich-toned, flute-like notes and yodelling phrasing. The male is all-black with a bright orange-yellow beak, the female dark brown above, lighter beneath. The blackbird feeds on the ground and frequently nests in hedges.

The **dunnock** is also known as the 'hedge sparrow', but it is not related to the sparrows; it has the needle-sharp bill of an insect-eater. Shy, almost furtive, brown and dark grey in colour, it jerks and hops about among the lower branches or underneath shrubs, and in and around the hedgerow plants, searching for small insects in summer and weed seeds in winter. While doing so the wings and tail are continually flicked. The male overcomes his shyness to sing a warbling song from the top of a low bush to his mate as she incubates their blue eggs in the delicate nest of grass, moss and rootlets, lined with wool, hair and feathers built in the hedgrow or bush. They prefer a nesting site that is evergreen or overgrown with ivy.

Dunnock (*Prunella modularis*)

The **wren** has been known as the mouse bird because of its small size, plumage colour (reddish brown with bar markings) and quick, restless movements as it busies about in the bottom of the hedgerow and on the ground. No mouse, however, has that comical nearly-always-cocked stumpy tail! Using its rapidly moving wings it can sustain flight across wide spaces, as across a road from one hedgerow to another, being seen by the surprised passerby for only a few seconds. It also has a habit of suddenly popping out on to a prominent twig, jerking up and down, then as suddenly disappearing again back in the hedgerow. Another surprise is the trilling song, heard all year, piercingly loud for such a small bird, the call is a loud 'tit-tit-tit' which becomes a churring noise if the wren is alarmed. The dome-shaped nest, with a side entrance, is built in a thick hedgerow, in ivy on a tree or among brambles, and made of moss, grass, leaves, fern fronds, lichen, etc, lined with feathers and hairs. The outside of the nest is camouflaged with material from nearby.

The **blue tit**, also known as the tom tit, blue cap and (in Scotland) blue bonnet, is the acrobat of the hedgerow and wayside. Conspicuous with its bright blue crown, wings and tail, and yellow underparts, it is to be seen high up in the branches of a tree actively searching here and there among the twigs, foliage and bark, first upside-down, then right way up, hanging in almost any position that allows it to snap up a caterpillar, spider, fly or other insect; or it may be low down in a hedgerow bush, rose thicket or bramble clump, again taking prey at an amazing speed. The flight is generally very fast over a short distance between trees or hedgerows. It nests almost anywhere it can find a suitable hole, in a tree, bank or wall, the nest being of moss and grass with a compacted lining of hair, wool and feathers. Incredibly, up to sixteen eggs may be laid in this nest and the tiny young successfully reared, but six to ten is more usual. (The great tit is not only larger but has a blackish head and neck and a black stripe down its front.)

The **whitethroat** is a summer visitor, arriving from Africa in late April to mid-May, departing any time from August to October. It prefers old, neglected hedgerows, field margins or waysides with clumps of furze and beds of nettles. A jaunty little bird with rusty-coloured wings and a long tail with white outer feathers and pinkish underparts, its song is a continual cheerful chatter from the living-place; but if the passerby approaches close to the nest it makes a harsh repeated 'chek, chek' note. A country name for it is the nettle creeper. In fact it builds its nest near the ground among nettles and other plants alongside the hedgerow or bushes, using grass with a lining of finer grasses and hair. Sometimes the female decorates the nest with bits of wool. Like other ground-nesting birds, its eggs or young can fall victim to snakes or rodents; if the male is heard making the repeated warning notes and restlessly flies about in the hedgerow or shrub, it is probably aware that some predator is on the prowl in the area of the nest.

Whitethroat (*Sylvia communis*)

The **blackcap** is a lover of tangled hedgerows, coppices and waste-places where brambles and honeysuckle abound. It is not seen so often as the whitethroat but is immediately recognisable, being sober grey with a black crown or cap (the female's is brown.) It arrives in April as a summer visitor and sings well into July, when other songsters, with their nestlings to feed, have fallen silent. The song is ended with a flute-like 'hee-ti-weeto-weeto', but the blackcap also steals the notes of other birds and has been known to mimic the nightingale and the garden warbler. The frail nest of dry grass, hair and fibre is built towards the base of a tangly rose or bramble. In recent years numbers of blackcaps have been overwintering in southern England and not leaving Britain in the autumn. Their diet is mainly insects, but they will also take soft berries like those on ivy.

Two handsome but very different black-and-white birds are often to be seen busying round the hedgerow. The **pied wagtail** is immediately recognised by the constant bobbing up and down of its long tail, black with white outer feathers, as it searches for insects on the ground. It is a familiar sight on waysides near gardens or fields, even in suburban areas.

The **magpie** belongs to the crow family and is a substantial bird, unmistakable in its smart black-and-white plumage as it hops rapidly across a road from hedge to hedge, walks along or chatters harshly from a tree or tall shrub. In sunshine its wings gleam with irridescent blues and greens and its tail has a purplish gloss. Magpie nests, domed and made of sticks, are often plainly visible in tall hedgerow trees. Not particularly nervous of people (or cars), magpies seem to have adjusted well to urban areas where biggish trees can be found. They eat a wide range of food — insects, snails, small rodents and — for which in the past they were persecuted by gamekeepers — the eggs and young of smaller birds, as well as fruit and vegetable matter. They will sometimes bury acorns as a squirrel does.

Related to the magpie, the **rook** can be identified from the carrion crow by a whitish patch around the base of its bill. It is also gregarious, moving about in flocks, whereas the carrion crow remains solitary or in a pair. Where the hedgerow trees are high, rooks will use them for a rookery, particularly if near open fields where there will be a food supply of beetle grubs, worms, leatherjackets, weed and crop seeds. Rooks do take corn and will damage other crops but this is usually outweighed by the pests they take in large numbers. The bulky nest is built of twigs and turf, cemented together with earth and lined with straw, grass, moss, hair and wool — it may be last year's repaired with fresh material or a completely new nest. Rooks will return year after year to the same rookery, then in one nesting season, for some unexplained reason, they will abandon it and choose a new site. It is said that if they build high it will be a fine, calm summer, but if they build lower down it will be a wet, windy summer!

The **kestrel**, our commonest hawk, is an interesting bird to see. It is chestnut brown, the male having a grey head and tail with a black tail band. Formerly it lived on open moorland, hillside, field or marsh. But it has adapted to hovering over the embankments and reservations along our motorways, finding them a refuge for the small rodents on which it feeds. No one tramps about there, and the traffic hurtling past on either side does not worry them. The kestrel has also learnt that changes in agricultural cultivation have forced some of the prey it used to find in the fields into the hedgerow verges. It is also known as a windhover due to its distinctive way of hanging stationary, head to wind, rapidly fanning its long pointed wings to keep itself in position: this is how you are most likely to see it from the car. If it cannot spot anything edible it quickly glides away to another position, repeating the operation again and again, until it finds an unsuspecting rodent or a beetle, on which it swoops with lightning speed by shutting its wings and dropping to earth. It uses the old nest of a crow or magpie, or a hollow tree in a hedge or woodland, a cavity in a ruin, a quarry or cliff face, in which to nest. It is also increasingly coming into towns and cities to hunt the park and garden hedgerows and nest on building ledges.

5
Mammals and reptiles

You may walk along the length of a hedgerow or wayside and not see a single animal, but that does not mean none are there. As you are moving, they sit tight and stay hidden, waiting for you, an unknown danger, to pass, or they sleep quietly until night time. To see the local animals and other creatures active in daytime, find somewhere secluded, behind a hedgerow tree or in an evergreen bush, even just sitting unobtrusively at the bottom of a stile, and keep still, quiet and watchful.

It may not be too long before you hear leaves being rustled and searched, then see foliage being pushed through or climbed up or over, or movement among the shrub branches or around the bottom of a tree trunk. If you do, remember not to blink; do it slowly if you must — a keen-sighted animal will see the movement of your eyelids and the flash of the whites of your eyes, and will be gone, at least for a while.

You may already have found evidence of animals being in the area. Where rabbits, foxes or badgers regularly pass through the hedgerow bottom they make a gap by pushing aside the vegetation, and the earth either side of this may be trodden hard, perhaps with footprints still visible. There may be other regularly used trackways in the hedgerow bottom.

Another sign is a near-bare patch where pellet-like rabbit droppings are grouped together, or among the debris the corkscrew-like pale brown droppings of a weasel or the dark brown, pointed, slug-like droppings of a hedgehog. Chewed or shredded grass and plant leaves, nibbled acorns, rose hips and hawthorn haws, hazel nuts with a hole in and the kernel gone, teeth marks on fallen crab apples and other wild fruit, sweet chestnuts half-eaten and with the rind stripped off, gnawed pine cones, toadstools with chunks gone from the cap or stem, young twigs with their tender bark gnawed, and the remains of dead birds and animals, of slugs or snails and their broken shells, broken birds' eggs, hair caught on spiny plants and shrubs: all these are evidence that animals are around.

Earth at the bottom of a hedgerow is often looser than in neighbouring land, and if it has a ditch it is also better drained. These two facts, with the protective cover of the hedgerow and close proximity of a regular supply of food, make it an ideal place for animals to burrow and make their summer and winter living accommodation and store place. Here you may find the entrance hole or holes of a rabbit colony, with earth scattered around the entrances. Smaller holes can be the entrance to a vole's or mouse's burrow; only by witnessing the animal using it or by identifying the tracks is it usually possible to identify the owner. A larger hole or holes, especially if a refuse dump of waste food, apples, etc, is heaped near or in the hedgerow, may be the entrance to the common brown rat's burrows.

You may see a young weasel or inexperienced rabbit dash across a road from one hedgerow to another. There may also be sad evidence upon the thoroughfare itself of animals that did not run fast enough. The rush across of a rabbit, vole, shrew, rat, mouse, stoat, weasel, snake or slow-worm, the slower perambulation of hedgehog, mole or toad, is so often on our traffic-busy country thoroughfares the final action before the creature's life ends as a flattened corpse.

The **hedgehog** is probably the most run-over animal in Britain and yet its population total shows no sign of declining. It lives not only in hedgerow bottoms, but also in woodlands, ditches, city parks or town gardens. Mainly nocturnal, it emerges at dusk from its daytime hiding place in a heap of leaves or debris to forage for its prey — slugs, snails, beetles, earthworms, insects, mice, rats, frogs, lizards, snakes, berries or, near dwellings, household waste such as bacon rind and meat scraps. It will come out after a heavy rainstorm to snap up the slugs and snails that also emerge. Never kill a hedgehog living in your garden: it consumes a vast number of slugs. In the past people used to keep a hedgehog or two in their kitchen to keep down the cockroaches, but — even if you have cockroaches — this is not advisable as hedgehogs often carry fleas. In the past, gipsies and romanies used to capture and kill hedgehogs, wrap them in a ball of clay and cook them in the hot ashes of a fire. When ready, the clay was broken off and with it came the 'spines' so the flesh could be eaten. I am told it tastes like pork, hence its rural name of hedgepig, but that at certain times of the year it is rather fatty! The male and female 'hedgepig' are called the boar and sow, apt names when one listens to them squeaking and grunting as they prowl around. They also snore while asleep. The 'spines' are really modified hairs which can be raised or lowered at will. Hedgehogs

Hedgehog (*Erinaceus europaeus*)

are strong climbers but if they fall will immediately roll into a ball so that the muscles around the base of the 'spines' take the shock of hitting the ground. Apart from motor vehicles their chief enemies are the fox and badger, but these only attack if acutely hungry. If a hardy fox does try to bite into the prickly ball, the hedgehog emits a noxious smell.

It hibernates in autumn, making a nest of dry leaves, moss and similar debris in a hedgerow bottom or a compost heap, under a pile of cut brushwood (one did so under my garden shed), or even inside an old wasps' nest enlarged and lined with dry materials.

The **weasel**, or its near relative the stoat, will occur wherever there are rabbits and small mammals such as mice to prey on in a hedgerow or wayside. Its short legs and lean, slender body allow it speedily to pursue a victim through the entrance hole and into the burrow to kill it underground. It is an agile climber, seeking birds' nests for young and eggs, it can leap what must seem to it wide distances, and it swims to pursue frogs or water voles or to escape.

Inquisitive and cunning, if disturbed it may dash into the vegetation, but if the passerby stands still, the weasel's head may soon pop out of its hiding-place, its steely-black eyes surveying the situation. Victims are killed by a bite at the base of the skull.

The weasel closely resembles the stoat, but is smaller and its tail does not have a black tip. It has reddish-brown fur with white underparts and runs with an arched back and a gliding movement. Despite its small size (about 10 inches with a 2 inch tail) it will courageously attack animals larger than itself, even rats that are equally fierce when cornered.

A **common shrew** sometimes gives away its presence in a hedgerow bottom by making high, shrill squeaks, particularly when it meets another shrew. The males are fierce and quarrelsome when confronting each other. Brown and grey in colour, the head and body are about 3 inches with a tail half as much again. Active by day and night, the long sensitive snout that extends beyond its mouth twitches as it continually searches for earthworms, insects, snails and slugs; the shrew has a hyperactive digestion — in 36 hours it consumes four times its own weight and if for some reason, perhaps through being trapped, it cannot obtain food, it starves to death in a very short time. Its bones are frail and though it climbs rapidly a shrew often dies from the impact of a fall. If picked up it can also die of shock. Although able to bury itself in loose earth in 12 seconds when surprised,

Weasel (*Mustela nivalis*)

this is rarely quick enough to prevent it from being seized by cats, hawks, owls, magpies, stoats, weasels or vipers. It spends the winter foraging in the hedgerow bottom and ditch, in summer moving out to the wayside and rough pasture where there is deep grass to shelter it.

The related pigmy shrew, Britain's smallest mammal, head and body 2¼ inches long, weight 5 grammes — less than a fifth of an ounce — also occurs in the same places and has similar habits. If you find a dead specimen a clue to its identity is that it has red tips to its teeth.

Common shrew (*Sorex araneus*)

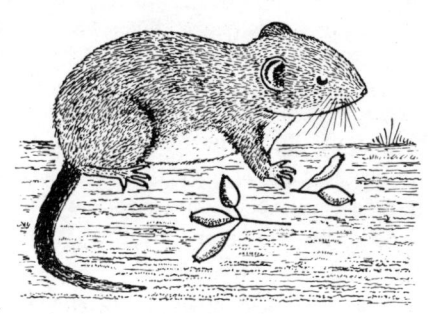

The **bank vole** inhabits hedgerows and woody landscape. It is bright, chestnut red with whitish underparts and long hairy tail; it measures about 3¾ inches. In daytime all the year round it climbs up among the twigs in search of berries, fruits and seeds, nuts, buds or the bark of young trees and shrubs. Nibbled hips and haws, either still on the plant or lying at the base, or bitten fragments of a toadstool indicate that bank voles are about. They dig up roots and bluebell bulbs, take wheat and barley grain from adjoining fields, take birds' eggs and young, snap up insects. They will kill other rodents and sometimes the quarrelsome males fight to the death, the victor eating the loser! Shallow burrows are made in a grassy hedgerow bank, in a sunny position, with many entrances and exits; the vole can dash down a hole at the bank top and reappear at one lower down some way away.

Bank vole (*Clethrionomys glareolus*)

The **long-tailed field mouse** or **wood mouse** (one and the same animal) commonly occurs in the hedgerow and wayside, where it burrows, and in gardens too. In the latter this attractive little creature can occasionally be a nuisance, digging up newly sown peas and bulbs, damaging or removing strawberries, gooseberries, apples or flower buds: it may also eat berries, grain, acorns, seeds, nuts, snails and insects. When disturbed it runs rapidly and can make prodigious zig-zag leaps to escape. In appearance it is similar to a house mouse, for which it is often mistaken, but it has more reddish-brown fur and

43

Long-tailed field mouse (*Apodemus sylvaticus*)

larger eyes and ears. Head and body are about 3½ inches, with a tail as long again. It is also a prolific breeder, but its numbers are kept under control by the many creatures that prey on it: owls, hawks, stoats, weasels, foxes, hedgehogs and vipers. It is so short-sighted that it can sometimes be quietly approached and captured under a hand.

The **harvest mouse** was originally mainly a cornfield dweller, but the use of the combine harvester drastically reduced its numbers. Survivors took up residence in suitable untrimmed hedgerows adjoining fields. Here they construct their ball-shaped summer homes, 3 inches in diameter, beautifully made of plaited and woven grass blades, among tall grass and vegetation, in bramble shrubs or furze bushes above the ground, insead of as formerly upon several stems of corn; by this adaptation it has not only survived but is beginning to increase. Charming to look at, timid in

Harvest mouse (*Micromys minutus*)

character, the tiny harvest mouse is active in the day, easily climbing up a stalk to sit on an ear of corn or head of cow-parsley seed. Head and body are about 2½ inches, with a tail as long again. The tail (scaly and almost naked) is so pliable that it can be coiled to grip a corn or grass stem almost as an extra foot. It constructs burrows in the hedgerow bottom to store grain for use in winter and here it spends periods asleep. In summer it mainly eats soft leaves and insects, grain, seeds and berries when available.

The **viper** or adder likes to lie in the sun to warm itself and prefers a sunny bank that has a hedgerow with a ditch nearby. It eats mainly shrews, voles and mice so the hedgerow provides it too with its food. Vipers do have a poisonous bite being the only British snake to have this, but they are given an undeservedly bad reputation. If they are poked with a stick or someone attempts to pick them up they will retaliate by biting, but normally they usually move away into the nearest cover, a nettlebed or bramble bush, even bracken, if they detect the vibrations of a human footfall. They are shy, retiring creatures, and in any case their mouth-gape is small, the danger is that they could seize a finger or toe, or the soft flesh of a hand or foot, if someone gathering low flowers such as primroses accidentally touched them. During autumn several vipers may be seen in the same hedgerow bottom, seeking a dry hole in the bank, a disused animal burrow or a hollow under a litter of debris and leaves in which to hibernate during winter. Colour and markings are very variable but there is usually a dark wavy or zig-zag line down the back, with spots and white dots on either side.

Viper (*Vipera berus*)

The **slow worm** is not a worm or a snake, but a harmless legless lizard. Its body is scaled like a snake's, but the anatomy has vestiges of limbs discarded some time during its centuries of evolution; it also has a notched tongue like a lizard, not the snake's forked tongue. It is golden yellow and bronze in colour, and very glossy. Its snake like appearance often leads ignorant people to kill it, but its diet is simply small earthworms, insects and spiders, and certainly no gardener should kill it because it consumes as many slugs as it can find. In spring it likes to doze in the sunshine on a warm hedgebank or in a sunny ditch, unless searching by day for prey in the hedgerow. Later in the year it spends all day in a hedgerow bottom where there are stones under which it can hide, coming out at dusk to feed. In winter it hibernates in an underground burrow or a hollow under a large stone. When alarmed it will dive into loose soil or beneath a stone and speedily vanish. Its main enemies, apart from man, are hedgehogs and vipers.

Another sunlover by the side of a hedgerow is the **common lizard** — not all that common today. It finds a bare patch of earth or a tree stump, and basks upon it, if disturbed shooting forward horizontally into the nearest cover. It moves rapidly over plants with a gliding motion, body and tail hardly lifted off the ground, to seize flies, beetles, moths, spiders and caterpillars; these it swallows whole unless too large, when it chews them for a while. It is difficult to capture, and if you succeed do not hold a lizard by its tail: it can snap this at a weak point and you will be left holding the wriggling end while the owner escapes. It will grow a new tail, shorter and blunter. After a lizard has been disturbed and dashed into the nearest hiding place, it will return to the sunny place when it thinks danger has gone. Lizards are supposed to be responsive to unusual sounds, and loud whistling is claimed to coax them back from hiding!

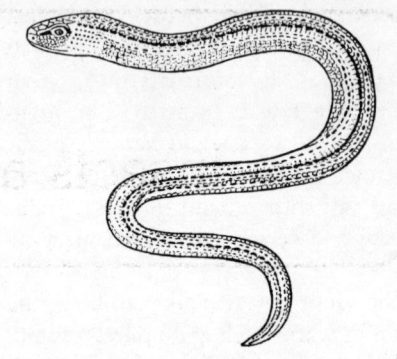

Slow worm (*Anguis fragilis*)

Common lizard (*Lacerta vivipara*)

6
Insects and others

The insects found in and along a hedgerow and wayside are of course those that can find there the protection they require, the trees and plants on which they lay their eggs, the plants or other insects on which they feed. A hedgerow with a wide range of flowering plants and shrubs is going to attract and house a larger number of different insects than one with only a small range of varieties. As with birds, many of the winged insects seen along a hedgerow or a wayside will also travel elsewhere, going to adjoining habitats such as woodland edges or fields. Some regularly occur along the hedgerow and wayside, a few colourful butterflies being particular favourites.

The **orange-tip** butterfly appears in April to June, in dancing flight from flower to flower. The male has the orange patch on the front wings, but the female has only a blackish-grey patch. The underside of the back wings of both male and female is beautifully marked with green mottling which shows through to the upper side. Eggs are laid on jack-by-the-hedge (Chapter 3), where the bluish-green caterpillars, with a white stripe on the side, may be seen.

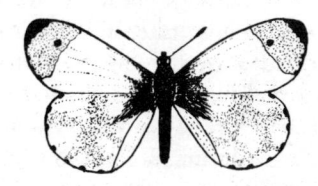

Orange-tip butterfly (*Anthocharis cardamines*)

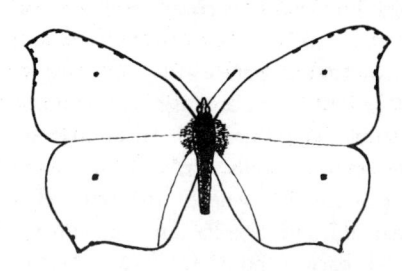

Brimstone butterfly (*Gonepteryx rhamni*)

The **brimstone** butterfly comes out even earlier, being attracted out of hibernation from February onwards if the weather is warmly favourable. The male is sulphur-yellow, the female a paler greenish-yellow; both have a central orange spot on each wing. Eggs are laid on buckthorn, the caterpillars being green in colour with black dots. It is claimed the name 'butterfly' was first used for the brimstone, which centuries ago was called 'the butter-coloured fly'; the name was later shortened and applied to all 'butterflies'.

The **small tortoiseshell** butterfly is common wherever there are beds of stinging nettles in a hedgerow or wayside. It is reddish orange with yellow patches, black-and-white spots and blue crescents, though markings are very variable. The green eggs are laid in a group on the underside of a nettle leaf in May. From these the black-and-yellow caterpillars emerge to feed as a colony on the nettles, using strands of silk to draw leaves together to form a shelter. When the leaves are consumed right down to their main ribs, the caterpillars establish another communal feeding place, and so on until half-grown; then they separate to feed singly. When there are enough of them the nettles can be stripped, although new shoots eventually grow. The small tortoiseshell is one of the butterflies that fly indoors in autumn, into a shed, barn or house, to find a place to hibernate; they will cling to the folds of a curtain or an old coat, or stay in a cupboard or dark corner for months on end.

Dragonflies are often seen zooming along hedgerow tops and edges and over waysides. These brilliant insects of many colours and patterns are great travellers, with such a speed of flight that they are found far from the lake or pond where they spent the early part of their life cycle in the water, in

Small tortoiseshell butterfly (*Aglais urticae*)

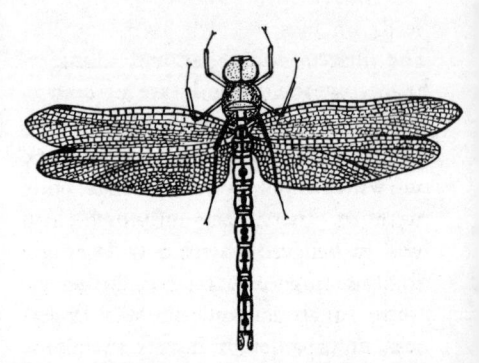

Dragonfly (*Odonata*)

Bumble bee (*Bombus*)

the 'nymph' stage. Long before man could do it dragonflies had mastered the act of hovering in flight, even flying backwards. The very large eyes have between 10,000 and 30,000 lenses, depending on the type of dragonfly. A dragonfly cannot focus its eyes as we do, but each lens looks in a different direction and so it can see movement from every angle. It feeds on insects, such as mosquitoes and flies, many of them pests, seizing them in flight, then perching on a twig and holding the victim with its front pair of legs while it severs the wings and pulps the victim's body with its jaws. Dragonflies are still widely believed to sting — they are sometimes known as 'horse stingers' — but they do not even possess any stinging apparatus. A more harmless insect, except to their insect prey, would be hard to find.

A **bumble bee** seen flying in and out of the low plants at the hedgerow bottom is probably a female searching for a suitable hole or disused animal-burrow entrance in which to make a nest. With her jaws she cuts grass blades, moss and bits of hay, which she takes into the hole and there creates a sort of matting. In the centre of it she makes a space, then flies in and out fetching flower pollen and nectar. In the nest centre she puts the pollen on the floor, then moulds it with nectar to make a paste ('bee bread') that she spreads in a circle or plate. On this she lays her eggs, covering them with a wall and roof of wax from special glands on her abdomen.

Using more wax she makes a 'honeypot' about the size of a hazel nut at the entrance to the nest, and puts in it a supply of honey for herself and her brood in bad weather. Then she sits on the wax 'roof', the warmth of her body helping to hatch the eggs. The grubs find a ready supply of food, which she adds to as needed from her honeypot. Within a week the grubs are full-grown; each makes itself a cocoon and changes to a pupa; in another fortnight they become adult bumble bees and fly away. In the meantime the mother bumble bee has made new paste and laid more eggs, so a further generation of adults is raised.

The name really should be humble bee, coming from the German *hummel* bee, referring to their humming noise; but it was thought that the bees 'bumble' along in their heavy flight, so the name bumble bee has stuck. There are many different sorts, some very handsome, some identified by the colour of their hairs on the largest part of the body: red-tailed bumble bee, yellow-tailed bumble bee, buff-tailed bumble bee among others.

The **common wasp** has spoilt many a wayside picnic: it has a remarkable ability to detect sweet substances, your jam sandwiches or sweets being especially attractive, as is flower nectar or ripe and over-ripe fruit — anything with a high sugar content. As the wasp has a narrow mouth, it can only lick up liquid foods. It will, however, grab other insects, such as grasshoppers and similar creatures,

49

paralyse them with a sting, cut off the wings, roll the remaining corpse into a ball, tuck this under its body, grasping it with its legs, then fly off to the nest to feed the victim to its grubs.

Common wasps will not deliberately attack people unless someone starts swatting at them, when they become angry and attempt reprisals. Most stings occur in late summer or autumn when wasps become 'sleepy' and may be touched or trodden on in the house. When a number of wasps are continually flying on to your picnic feast, walk back along both sides of the hedgerow for a considerable distance: if you see wasps frequently coming in and out of a hole in the hedgerow bottom or hedgebank and returning to it, then that is the site of their nest colony, which may house up to 25,000 wasps in a season! They are not likely to attack if you just stand still watching them, because they are far too busy, but it is unnerving to share your chocolate cake with these seemingly aggressive insects, so beat a retreat to a further part of the hedgerow or wayside. Badgers have no such fears, being equipped with a thick fur coat that angry wasps will not sting through. If you see the torn, grey, paperlike remains of a wasps' nest pulled out of the hedgerow bottom, this is where a badger or two obtained a wasp-grub meal.

The **crab spider** is just one of several spiders common among the hedgerow and wayside vegetation. Others create

Common wasp (*Vespa vulgaris*)

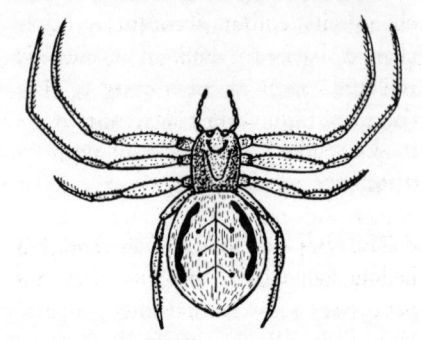

Crab spider (*Misumena calycina*)

the well-known orb cobwebs, complete with trip wires that cause passing insects to fall into the trap. The crab spider, though, does not spin any sort of web. It just waits for a victim to come to it. After climbing a plant or shrub, moving its squat body with a crablike action, it may crawl inside an open flower, such as bramble, pressing itself against the central parts of the flower or the petals to make itself unspiderlike in shape; its first two pairs of long, powerful forelegs outstretched and waiting. When a bee, butterfly or fly alights on the flower, the spider's powerful legs have seized it before it has realised its mistake, and it is soon a corpse being sucked out by the spider. If you see a butterfly or other insect struggling upon a flower a